# 5th International Conference on Plant Synthetic Biology, Bioengineering & Biotechnology 2021



Online
15-17 November 2021

ISBN: 978-1-7138-4529-4

**Printed from e-media with permission by:**

Curran Associates, Inc.
57 Morehouse Lane
Red Hook, NY 12571

**Some format issues inherent in the e-media version may also appear in this print version.**



Printed with permission by Curran Associates, Inc. (2022)

For permission requests, please contact AIChE
at the address below.

AIChE
120 Wall Street, FL 23
New York, NY 10005-4020

Phone: (800) 242-4363
Fax: (203) 775-5177

www.aiche.org

**Additional copies of this publication are available from:**

Curran Associates, Inc.
57 Morehouse Lane
Red Hook, NY 12571 USA
Phone: 845-758-0400
Fax: 845-758-2633
Email: curran@proceedings.com
Web: www.proceedings.com

# TABLE OF CONTENTS

## ORAL SESSION

## POSTER SESSION

**A Previously Unknown Biosynthetic Capacity of Chalcone Synthase Discovered By Heterologous Expression of a Putative Plant Biosynthetic Gene Cluster in Yeast**

Authors:

Li, S. - Presenter, Cornell University

Smolke, C. D., Stanford University

Kong, D., Stanford University

Plant biosynthetic gene clusters (BGCs) are operon-like organization pattern of genes in plant genome. Although BGCs are less abundant in plants than in microorganisms, more than 30 plant biosynthetic pathways, mainly in plant specialized metabolism, have been identified as BGCs in the past decades from diverse plants. As with the successful application of microbial BGCs in discovering novel microbial biosynthetic pathway, plant BGCs have been a promising approach to accelerate novel plant biosynthetic pathway and plant natural product discovery. For example, investigation of the known plant BGCs have led to the discovery of unexpected, novel pathway components (e.g., sugar transferases, transporters, and regulators) that cluster together with known pathway genes. However, the discovery of novel biosynthetic pathways and biosynthetic capabilities from putative BGCs is still in its infancy.

Here, we firstly identified a previously unknown amide biosynthetic pathway derived from tomato (Solanum lycopersicum) from predicted tomato BGC candidates. By expressing a putative tomato gene cluster in yeast (*Saccharomyces cerevisiae*), we identified two functionally related enzymes catalyzing cascaded amide biosynthetic reactions. The newly discovered pathway consists of a carboxyl methyltransferase (SlMT2), which methylates the primary metabolite 3-hydroxyanthranilic acid (3-HAA) to form a methyl ester, and a naringenin chalcone synthase (SlCHS), which catalyzes the condensation of 3-HAA methyl ester and p-coumaroylâ€“coenzyme A (CoA) through formation of an amide bond. The newly discovery nitrogen-carbon bond formation biocatalytic capacity of CHS is different from its well-characterized Polyketide Synthase activity that canonically catalyzes carbon-carbon bond formation through iterative decarboxylative Claisen condensation. We further demonstrated that this aminoacylation activity could be a common secondary activity in plant CHSs by validating the activity in vitro with variants from *S. lycopersicum* and *Arabidopsis thaliana*. Our work demonstrates yeast as a platform for characterizing putative plant gene clusters with the potential for compound structure and enzymatic activity discovery.

**Accelerated Discovery and Engineering of Medicinal Plant Alkaloid Biosynthetic Pathways**

Authors:

Nett, R. - Presenter

Sattely, E., Stanford University

Plants have served as crucial medicines for thousands of years, and continue to supply us with clinical pharmaceuticals and novel compounds with diverse bioactivities. However, our ability to access the molecules produced by medicinal plants is often limited by their slow growth, difficulty in cultivation, and the threat of over-harvesting native populations. One potential strategy to alleviate these challenges is through the metabolic engineering of plant biosynthetic pathways for medicinal compounds, but this requires underlying knowledge of the requisite metabolic genes, as well as a heterologous system in which pathways can be rapidly tested and assembled. By leveraging the conditional production of plant-derived molecules in native plants, we have been able to utilize a combined metabolomic and transcriptomic approach for the rapid discovery of biosynthetic pathways for multiple medicinal plant alkaloids. This includes a near complete pathway for huperzine A, a psychoactive compound with potential as a treatment for neurological diseases, as well as a full biosynthetic pathway for colchicine, an ancient and still-utilized treatment for inflammatory conditions. Critically, we have not only established a full set of genes for colchicine biosynthesis, but we have also engineered its complete pathway, consisting of 20 genes, into the model plant *Nicotiana benthamiana*, thereby demonstrating an alternative means to produce this pharmaceutical molecule. Our results highlight a powerful strategy for the discovery of plant biosynthetic pathways, and further provide a proof-of-concept for the engineered production of medicinal alkaloids. Additionally, our insight into the biosynthetic reactions used to produce these molecules has led to the discovery of novel catalytic mechanisms for well-known enzyme families, as well as the identification of new classes of specialized metabolic enzymes, thereby expanding our understanding of the remarkable breadth and diversity of chemistry used within plant metabolism.

**Bioactive Compounds from Extracts of Sarcopoterium Spinosum**

Authors:

Erdem, C. - Presenter, Ege University

Bayraktar, O.

Herbal medicines have traditionally been used to prevent and treat various diseases. *Sarcopoterium spinosum* (*S. spinosum*), was reported as a potential source of natural compounds to be used in the treatment or prevention of various health problems such as chronic pain, digestive problems, diabetes, and cancer (Hammad, 2018; Smirin et al., 2010). In this study, the antidiabetic effects of the extracts from the aerial and root parts of the *S. spinosum* were investigated using bioactivity-guided isolation methods. Extraction parameters (percentage of ethanol, extraction time, and solid to liquid ratio) were optimized based on the responses of total antioxidant capacity, total phenol content, and extraction yield, using the Design-Expert program. The extraction conditions for the highest activity of extracts were determined and the isolation of bioactive target compounds was carried using preparative thin-layer chromatography (Prep. TLC). The optimized conditions for extraction of root part were determined as 24 h extraction time, 1:20 solid-liquid ratio, and 90% ethanol. For the extraction of the aerial part, the total antioxidant capacity, total phenol content, and productivity were at their maximum values at extraction conditions of 1:30 solid-liquid ratio, 80% ethanol, and 4-hour. Alpha-glucosidase inhibition tests were performed for the bioactive target compounds obtained from Prep TLC. Acarbose was used as a positive control. Relatively higher inhibition percentages were obtained than acarbose for the two of the bioactive compounds isolated from the extract of the root part.

REFERENCES

Hammad, K.A. (2018). Chemical composition of the methanolic extract from seeds, thorns, and leaves of *Sarcopoterium spinosum* (L.) (Rosaceae) grown in Al-Tafila, Jordan. *Research on Crops*, 19, 315â€"319.

Smirin, P., Taler, D., Abitbol, G., Brutman-Barazani, T., Kerem, Z., Sampson, S.R., Rosenzweig, T. (2010). *Sarcopoterium spinosum* extract as an antidiabetic agent: In vitro and in vivo study. *Journal of Ethnopharmacology*, 129, 10â€"17.

**Carrier Gas Triggered Biolistic Delivery of Genetic Material with ZIF-8**

Authors:

Wijesundara, Y. - Presenter

Biolistic transfectionâ€"physically shooting DNA loaded onto a metal nanoparticle into a cellâ€" is an established method for introducing genetic material into cells that are otherwise difficult to transfect using traditional methods. Commercial gene guns and the gold or tungsten microparticle â€œbulletsâ€ that carry the DNA into the cells are expensive and there are concerns about their cytotoxicity, biopersistence, and potential for DNA degradation with the genetic material placed on the microparticle surface. We have developed an approach that biomimetically encapsulated DNA in Zeolitic Imidazolate Framework-8 (ZIF-8) as an efficient, low cost, and stable gene carrier that can be used with a low-cost gas powered â€œMOF-Gunâ€ for direct delivery into plant tissue. ZIF-8 is more economical, biodegradable, and affords protection to the DNA against degradation compared to metal nanoparticles. By using $CO_2$ as a reactive carrier gas, we can cause a rapid local pH drop and fast release of the DNA.

**Continuous Directed Evolution of Short-Lived Plant Enzymes for Reduced Maintenance Respiration Costs**

Authors:

Hanson, A. - Presenter, University of Florida

To ensure adequate nutrition for a fast-growing world population, overall production of major food crops must rise substantially. Conventional breeding may be too limited in potential and too slow to satisfy this rising demand, making it important to deploy also metabolic engineering and synthetic biology to increase crop biomass and harvested yields. One unexplored metabolic engineering strategy is to decrease carbon loss by slowing protein turnover (degradation and resynthesis) and hence reducing the maintenance component of respiration. The carbon saved thereby could be channeled into biomass. In plants, diverse enzyme are candidates for reduction in their turnover rate. An extreme example is the suicidal THI4 thiazole synthase that forms the thiazole moiety of thiamin (vitamin B1) and is an abundant protein in all plants. THI4 is not a true catalyst because one enzyme molecule is required per reaction cycle due to destruction of an active-site cysteine residue to provide the sulfur atom for the thiazole ring. Destruction of the cysteine irreversibly inactivates the enzyme. Based on CCR (â€˜catalytic cycles until replacementâ€™) values, which provide an estimate of an enzymeâ€™s energy cost in relation to its operation, we selected abundant, short-lived enzyme candidates in Arabidopsis. Besides suicide THI4s, the selected candidates include methionine synthase 2 and phosphomethylpyrimidine synthase as well as cereal THI4 isoforms that have no active-site cysteine and may therefore be truly catalytic instead of suidical. Using the yeast OrthoRep continuous directed evolution system, which has been successfully used to evolve microbial THI4s, we are evolving the selected candidates for longevity and improved performance. When used to replace the endogenous gene in its native host, the improved enzymes are expected to decrease maintenance respiration costs leading to increased biomass.

**Engineering of Phytosterol-Producing Yeast Platforms for Functional Reconstitution of Downstream Brassinosteroid Biosynthetic Pathways**

Authors:

Xu, S. - Presenter, UC Riverside

Li, Y., Stanford University

As essential structural molecules for plant plasma membranes, phytosterols are key intermediates for the synthesis of many downstream specialized metabolites of pharmaceutical or agricultural significance, such as brassinosteroids and withanolides. *Saccharomyces cerevisiae* has been widely used as an alternative producer for plant secondary metabolites. Establishment of heterologous sterol pathways in yeast, however, has been challenging due to either low efficiency or structural diversity, likely a result of crosstalk between the heterologous phytosterol and the endogenous ergosterol biosynthesis. In this study, we engineered campesterol production in yeast using plant enzymes and enhance the titer of campesterol by upregulating the mevalonate pathway. Further engineer to inactivate yeast acyltransferases enhanced the production of free campesterol. By employing metabolic engineering, strain evolution, fermentation engineering, and pathway reconstitution, we were able to establish a set of phytosterol-producing yeast strains and reconstitute and characterize the early brassinosteroids biosynthetic pathways on this phytosterol platform. In addition, the phytosterol-producing yeast platform were also used to reconstitute the plant membrane-anchored enzymes with enhanced enzymatic activities. This work resolves the technical bottlenecks in phytosterol-derived pathway reconstitution in the backerâ€™s yeast and opens up opportunities for efficient bioproduction and pathway elucidation of this group of phytochemicals.

**Engineering Rubisco for Food Security**

Authors:

Davis, R. - Presenter, Sandia National Laboratories

Sarwar, M., Sandia National Laboratories on behalf of the

Shinde, S., Sandia National Laboratories

Sarnaik, A., Arizona State University

Faisal, M.

Varman, A. M., Arizona State University

Global demand for high quality digestible protein is expected to substantially outpace population growth over the coming decades, necessitating gains in feed-to-food conversion efficiency for sustainable food security. Toward these ends, identifying and producing new high quality protein feedstocks that supplement existing agriculture will be of major importance and have significant impacts that intersect with the energy and water infrastructure. RuBisCo, the enzyme complex that initiates carbon fixation, is often cited as the most abundant soluble protein on earth, typically representing 40 - 60% of the total protein content of alga and green leafy biomass. In addition to its abundance, RuBisCo is easy to extract and highly digestible, as well as having other favorable chemical and mechanical properties for food and feed applications, including foam formation and gelation. Although RuBisCo has a very high native amino acid score (PDCAAS ~ 98/100), we have recently demonstrated significant improvements in the amino acid value by increasing expression of two key amino acids, methionine (Met) and lysine (Lys), which are typically limiting and costly for feed applications, in the large subunit of the RuBisCo complex, RbcL, of *Synechocystis* sp. PCC6803. Enrichment targets for Met and Lys in RbcL were identified using Grantham's distance matrix, with avoidance of previously reported lethal mutations. Installation of the selected codon swaps was achieved using Circular Polymerase Extension Cloning (CPEC), yielding between 60-100% higher Met and Lys abundance, respectively. Expression of Met- and Lys- enriched RbcL in *Synechocystis* sp. PCC6803 showed no detriment to specific growth rate or total quantity of RbcL, indicating technological feasibility for a sustainable solution to global protein deficiency. Similar engineering strategies should be amenable for terrestrial food crops and green leafy biomass.

**Establishment of Strigolactone-Producing Bacterium-Yeast Consortium**

Authors:

Li, Y. - Presenter, University of California, Riverside

Strigolactones (SLs) are a class of phytohormones playing diverse roles in plant growth and development, yet the limited access to SLs is largely impeding SL-based foundational investigations and applications. Here, we developed *Escherichia coliâ€"Saccharomyces cerevisiae* consortia to establish a microbial biosynthetic platform for the synthesis of various SLs, including carlactone, carlactonoic acid, 5-deoxystrigol (5DS; 6.65 Â± 1.71 Î¼g/liter), 4-deoxyorobanchol (3.46 Â± 0.28 Î¼g/liter), and orobanchol (OB; 19.36 Â± 5.20 Î¼g/liter). The SL-producing platform enabled us to conduct functional identification of CYP722Cs from various plants as either OB or 5DS synthase. It also allowed us to quantitatively compare known variants of plant SL biosynthetic enzymes in the microbial system. The titer of 5DS was further enhanced through pathway engineering to 47.3 Î¼g/liter. This work provides a unique platform for investigating SL biosynthesis and evolution and lays the foundation for developing SL microbial production process.

**Evaluation of an Algal Turf Scrubber (ATS) for Nitrogen Removal from High Strength Nitrate Wastewater**

Authors:

Terracciano, A. - Presenter

Abraham, J., Stevens Institute of Technology

Su, T. L., Stevens Institute of Technology, Center for Environmental Systems

Koutsospyros, A., Stevens Institute of Technology

Christodoulatos, C., Stevens Institute of Technology

Braida, W., Stevens Institute of Technology

Smolinski, B., US Army RDECOM-ARDEC

The production of plastics, nitric acid, nitrogen-based fertilizers as well as fuel and explosive components often results in the generation of wastewater volumes with high levels of inorganic nitrogen, often in the form of nitrate. The nitrate removal is a challenge of particular significance for traditional physical-chemical treatment technologies as the processes are often energy intensive with little ability to recover viable resources for further valorization of the original waste. In this scenario, microalgae treatment systems (MTSs) are gaining increasing attention as industrial ecology based technology; in fact, MTSs allows to achieve resource recovery in the form biomass revealing the inherent value of waste streams. Among MTSs, algal turf scrubber (ATS) represent a resilient and feasible technology with high efficacy for remediation of a variety of wastewater. ATS rely on the artificial composition of wastewaters to stimulate microalgal productivity, which leads to nutrients (C,P,N) assimilation into the biomass. In this study, an indoor bench scale ATS set up was inoculated with a filamentous algae consortium and utilized for treatment of high strength nitrate wastewater. Changes in the algae consortium were observed over time with the filamentous genus *Ulothrix* sp. rapidly becoming the dominant alga strain; the filamentous genus resulted in ease in harvesting and resistance to predation. Batch tests were conducted according to different nitrogen to phosphorous ratio (N:P) and recirculation rates. Nutrients uptake as well as biomass productivity were determined for each of the conditions tested and used to estimate the performance of the system at larger scale. Overall, the system was able to achieve up to 50% of the influent nitrogen load with biomass productivity in the range of 3-10 $g_{dryweight}$ $m^{-2}$ $d^{-1}$. In addition, the recirculation rate showed to have higher effect on the nitrogen removal as well as biomass production compared to the initial N:P ratio.

**Exploration of Epigenetic Regulation and Development of Global and Targeted Epigenetic Engineering Tools for Taxus Plant Cell Culture**

Authors:

Brzycki Newton, C. - Presenter, Worcester Polytechnic Institute

Microbes play critical roles in mediating the health of soil and plants. Synthetic biologists have developed a variety of production, sensing, and response capabilities in living microbes; however, these tools currently cannot be implemented in the environment due to the risk of uncontrolled proliferation of engineered microbes. To address this issue, my lab has been working on designing strategies for intrinsic biological containment of microbes, with a focus on the technique of synthetic auxotrophy. A synthetic auxotroph is an organism engineered to depend on the provision of synthetic nutrients not found in nature. This poster highlights our latest characterization of Escherichia coli synthetic auxotrophs, which includes continuous evolution and direct addition to mammalian cell cultures. It also describes our successful efforts to engineer the incorporation of synthetic amino acids in the gram-positive microbe Bacillus subtilis. B. subtilis is a model system for both the study of bacterial cell biology and for industrial uses as a probiotic, rhizobacterium, efficient protein secretor, and spore-former. To enable greater understanding and control of proteins in B. subtilis, we achieved incorporated many distinct synthetic amino acids within proteins. We used these systems to achieve click-labelling, photo-crosslinking, and translational titration. Synthetic auxotrophy in E. coli and eventually in B. subtilis could serve as an enabling safeguard for engineered plant probiotics, particularly those with pre-programmed limited lifetimes in the environment.

**Exploration of Epigenetic Regulation and Development of Global and Targeted Epigenetic Engineering Tools for Taxus Plant Cell Culture**

Authors:

Brzycki Newton, C. - Presenter, Worcester Polytechnic Institute

Roberts, S., University of Massachusetts, Amherst

Young, E., Worcester Polytechnic Institute

Mazzoni, G., Worcester Polytechnic Institute

Whittles, E., Worcester Polytechnic Institute

Gamache, M., Worcester Polytechnic Institute

Lenka, S., TERI Deakin Nanobiotechnology Centre

Many medicinal plants have developed unique biosynthetic capabilities that make production of specialized metabolites in plant cell culture (PCC) an attractive alternative to extensive engineering of model hosts. Yet, long-term continuous subculturing of PCCs can reduce productivity due to increased DNA methylation of promoters, transcription factors, and other regulatory elements, resulting in temporal downregulation of secondary metabolism. Here, we present our identification of epigenetic engineering targets and development of both global and targeted DNA methylation engineering tools for improving production of the chemotherapeutic drug paclitaxel (Taxol) in *Taxus* PCC.

We have demonstrated that treatment of *Taxus cuspidata* and *Taxus chinensis* PCCs with 5-azacytidine (5AC), a nonspecific DNA methyltransferase inhibitor, results in a 5 to 20-fold increase in accumulation of paclitaxel precursors 10-deacetylbaccatin (10-DAB) and baccatin III with no negative effects on cell viability or biomass accumulation. Yet, despite this increase in precursor production, treatment with 5AC had no significant effect on paclitaxel yield, indicating a need for a mechanistic understanding of epigenetic regulatory mechanisms and more targeted epigenetic engineering approaches to increase end-product yield. Next, differential transcriptome analysis of *Taxus cuspidata* PCCs in response to elicitation with methyl jasmonate (MeJA), a well-characterized activator of secondary metabolism in numerous plant systems, was used to identify potential gene targets for targeted epigenetic engineering. Studies revealed that phenylpropanoid metabolism (which competes for the paclitaxel precursor phenylalanine) was significantly upregulated under MeJA elicitation, which was confirmed by an observed increase in both lignin content and total flavonoids in MeJA-elicited cultures. Finally, we will discuss our development of CRISPR/dCas9-based tools for targeted demethylation of paclitaxel biosynthetic pathway genes and targeted methylation of phenylpropanoid metabolism genes. These approaches are easily generalizable and can be applied in parallel with traditional metabolic engineering methods to achieve stable, high productivity in a variety of different plant cell culture systems.

**Improving Catalytic Thi4s Using Continuous Directed Evolution Systems in Escherichia coli**

Authors:

Leong, B. - Presenter

Hanson, A., University of Florida

Continuous directed evolution accelerates enzyme improvement, enables deep exploration of the targetâ€™s fitness landscape, and is revolutionizing metabolic engineering. Several systems to continuously evolve enzymes are now available including OrthoRep, PACE, T7-DIVA, and eMutaT7. We are applying T7-DIVA and eMutaT7 to improve Thi4 thiazole synthases, which are thiamin synthesis enzymes. Canonical plant Thi4s are suicidal (i.e. self-inactivate after one reaction cycle) because they use an active site cysteine residue as the sulfur donor for the reaction. Certain prokaryotes have catalytic Thi4s that use sulfide as sulfur donor, but these organisms are mostly strict anaerobes or thermophiles from high-sulfide environments whose Thi4s are unlikely to function efficiently in plants. Our aim is to evolve catalytic Thi4s to operate under the plant-like conditions (aerobic, mild temperature, low sulfide) in *Escherichia coli*. Pilot data from T7-DIVA based on GFP expression (the test for transcription via the T7-RNAP/cytidine deaminase (T7RNAP-CD) fusion) showed this system functions in MOPS minimal medium. We complemented an *E. coli* Î”*thiG* thiazole auxotroph with the Thi4 from *Saccharicrinis fermentans* but found a severe growth penalty from the T7-DIVA machinery that blocked efforts to evolve this enzyme. The growth penalty may reflect downregulation of the target gene by the antisense orientation of the T7 promoter. Another concern is T7-DIVA mutagenesis machinery could potentially lose mutagenic activity since the RNAP driving mutagenesis is independent from the endogenous RNAP that drives target gene expression and imposes a fitness burden that may predispose it to loss. We also explored eMutaT7, a system that uses a similar T7RNAP-CD fusion, but ties target gene expression to transcription driven by the fusion protein. We showed that the eMutaT7 system functions in MOPS minimal media, and are now using eMutaT7 to proceed with Thi4 enzyme evolution experiments.

**Mechanism, Regulation, and Ecology of Bacterial Auxin Degradation By the Genus Variovorax in the Complex Root Microbiome**

Authors:

Conway, J. M. - Presenter

Dangl, J. L., University of North Carolina Chapel Hill

Walton, W. G., University of North Carolina

Redinbo, M. R., University of North Carolina

Salas-González, I., University of North Carolina Chapel Hill

Chemical signaling in the plant microbiome can have drastic effects on microbial community structure, and on host growth and development. Previously, we demonstrated that the auxin metabolic signal interference performed by the bacterial genus *Variovorax* via a novel auxin degradation locus was essential for maintaining stereotypic root development in an ecologically-relevant bacterial synthetic community. Here, we dissect the *Variovorax* auxin degradation locus to define the genes necessary and sufficient for indole-3-acetic acid (IAA) degradation and signal interference. We determine the crystal structures and binding properties of the locusâ€™s MarR-family repressor with IAA and other auxins. We identify auxin-degradation loci across the bacterial tree of life and define two distinct types based on gene content and metabolic products: *iac*-like and *Variovorax*-like. We solve the structures of MarRs from representatives of each auxin degradation locus type, establishing that each have distinct IAA binding pockets. Comparison of representative IAA degrading strains from diverse bacterial genera show that while all degrade IAA, only strains containing a *Variovorax*-like auxin degradation locus interfere with auxin signaling in a complex synthetic community context. This suggests that *Variovorax*-like locus containing strains play a key ecological role in modulating auxins in the plant microbiome.

**Mining Curated Plant Biosynthetic Gene Clusters in Coffea Spp Genomes**

Authors:

Correia de Lemos, S. - Presenter

Medema, M. H., Wageningen University

Domingues, D. S., UNESP

Paschoal, A. R., UTFPR-CP

Specialized metabolites from plants constitute the majority of natural products used as medicines and agrochemicals. Some of the enzymes responsible for specialized metabolite biosynthesis are encoded by genes that are proximally collocated in genomes - Biosynthetic Gene Clusters (BGCs). In coffee (*Coffea* spp), several compounds responsible for the complexity of flavor and aroma are specialized metabolites. Identifying homologous gene clusters aids in the study of their function and evolution. The MIBiG database is a repository containing curated BGC data and currently has 20 plant BGCs. Using BGC data from the MIBiG database and a preliminary inventory of putative BGCs in *Coffea* we assessed whether we could identify homologues of curated BGCs in the genomic sequences of three *Coffea* species: *C.arabica, C. canephora* and *C. eugenioides.* We used cblaster to perform searches using the protein sequences of each curated plant BGC as queries. Eight of the known plant BGCs had potentially homologous regions in *Coffea* species - one related to saccharide biosynthesis (dhurrin), one related to alkaloid biosynthesis (thebaine) and six are related to terpenoid biosynthesis. As a case study, we identified gene clusters partially homologous to the triterpenoid cucurbitacin BGC from *Cucumis sativus,* composed of six genes, including one cyclase, one acyltransferase and four P450s. In *C. arabica*, the partially similar gene cluster contains a putative acyltransferase and three P450s. In *C. eugenioides*, another similar genomic region was identified containing a putative acyltransferase and two P450s. These constitute interesting candidates for further investigation of similar metabolite compounds across different plant species. This approach is an important step in prioritizing BGCs for functional analysis and synthetic biology studies for the production of *Coffea* specialized compounds.

**Moontag: An Improved Programmable Transcription Activator for Plants**

Authors:

Casas Mollano, J. A. - Presenter

Zinselmeier, M. H., University of Minnesota Twin-Cities

Smanski, M. J., University of Minnesota, Twin Cities

CRISPR-Cas9 based transcriptional activators have been developed to induce gene expression in eukaryotic and prokaryotic organisms. The main advantages of CRISPR-Cas9 based systems is that they can achieve high levels of transcriptional activation and are very easy to program via pairing between the guide RNA and the DNA target strand. One of these CRISPR-Cas9 systems is SunTag, a second-generation system that activates transcription by recruiting multiple copies of an activation domain (AD) to its target promoters. SunTag has been tested in Arabidopsis showing to be a strong transcriptional activator. However, we found that in the monocot plant Setaria, SunTag was difficult to stably express in transgenic plants. To overcome this problem, we designed MoonTag, a novel activator that works on the same basic principle of SunTag, but whose components are well tolerated when stably expressed in transgenic plants. In Setaria, MoonTag is capable of inducing high levels of transcription of a reporter gene as well as of endogenous genes. More important, MoonTag components are expressed in transgenic plants to high levels without any deleterious effects. MoonTag is also able to efficiently activate genes in eudicotyledonous species such as Arabidopsis and tomato. We further increased the activation potential of MoonTag by replacing VP64, originally used in SunTag, by other ADs from plant origin. Thus, MoonTag represents a novel activator that could be used to regulate the transcription of endogenous genes in many plant species.

**Multilevel Modeling of Lignin Biosynthesis in Brachypodium distachyon**

Authors:

Barros, J. - Presenter, University of North Texas

Despite the potential of lignin as raw material for conversion into renewable bioproducts, our current understanding of its biosynthesis is still incomplete, especially in the economically important Poaceae family of grasses. Here, we applied a combined proteomics and isotope labeling approach to study lignification in a set of knockdown lines for several lignin biosynthetic genes using the model grass *Brachypodium distachyon*. We identified the lignin protein family members in mature stems, and found that PTAL, the first enzyme of the lignin pathway in grasses, was the fourth most abundant protein among the +11,000 proteins identified. Accordingly, p-coumarate, the product of the PTAL reaction, was the most abundant lignin pathway intermediate. Overall, our $^{13}C$-labeling data, metabolic flux analysis and in situ hybridization experiments support a metabolic separation of the carbon flux from L-phenylalanine and L-tyrosine into different lignin subunits and flavonoid classes associated with distinct cellular localization. Notably, co-downregulation of pairs of monolignol biosynthetic genes did not result in additive impacts on plant growth or lignin deposition, but instead induced pronounced proteome changes associated with oxidative stress and vitamin E synthesis. These results highlight the pivotal role of PTAL and the flexibility of the monolignol biosynthetic pathway in grasses.

**Translate this page**

**Plant Synthetic Biology Approaches to Decipher Solanaceae Acylsugar Structural Effects on Insect Herbivory**

Authors:

Lou, Y. R. - Presenter

Last, R., Michigan State University

Plants use their remarkable specialized metabolic capacity to interact with the biotic environment, including deterring and attracting insects and microbes. Solanaceae trichome-synthesized acylsugars confer defense against herbivores and offer the possibility of engineering plants with environmentally safe pesticides. Acylsugars show striking structural variety within Solanaceae family, with mono- and di-saccharide cores and acylchains of C2-C12 found across the family. We ask whether the repeated structure diversification bears the fingerprint of an evolutionary arms race. An alternative hypothesis is that the diversity is simply the result of genetic drift. Deciphering effects of structurally diverse acylsugars faces the challenge of interference of other defense traits (for example, different levels of alkaloids and terpenes in plants). To critically assess the impact of acylsugars on herbivory, we used CRISPR/Cas9 gene editing of domesticated *Solanum lycopersicum* and *Solanum pennellii* (LA0716) pathway enzymes to modify the amounts and types of metabolites produced. We generated isogenic *S. pennellii* lines differing in glucose- or sucrose-based esters to critically assess the impact of sugar core. Meanwhile, knockout lines of the first three acylsugar acyltransferase (ASAT) genes in both backgrounds accumulate no detectable acylsugars, without altering plant growth or trichome development. These lines provide a good 'null mutant' background to test the efficacy of acylsugars. Finally, cultivated tomato *slasat4* mutants lacking acetylated acylsucroses permit us to evaluate the impact of acetyl groups in insect protection. Our results suggest that even the performance of the specialist, *Manduca sexta*, is hindered by the presence of acylsugars, while we observe larger protective effects of acylsucroses than acylglucoses. The CRISPR mutated plants are suitable for investigating the importance of acylsugars in the field. Moreover, these plants serve as a promising molecular engineering platform that allows for the investigation of the mechanisms underpinning specialized metabolic evolution.

**Public Engagement with Biotechnology: A Scoping Review**

Authors:

Menary, J. - Presenter, University of Oxford

Fuller, S. S., University of Oxford

The responsible research and innovation (RRI) framework promotes the alignment of science with the needs, expectations and values of society. This is particularly relevant for research and innovation in technologies that have proven controversial, such as biotechnology in Europe.

We conducted a scoping review of research on public engagement with biotechnology to inform our work with PhotoBoost, a Horizon 2020 project that aims to improve the photosynthetic efficiency of key crop plants potatoes and rice through trans- and cis-genic breeding techniques.

We screened 2357 citations related to public engagement and biotechnology, retaining 37 studies and extracting data against a set of *a priori* and emergent indicators, as well as collecting methodological and demographic data; we then conducted a qualitative synthesis to explore the key themes arising from the existing literature.

Our key findings relate to the preponderance of WEIRD (*white, educated, industrialised, rich* and *democratic*) country research, a wide-range of engagement activities but reliance on focus groups and a number of key concerns for the public around the use of biotechnology, such as transparent weighing of risks, benefits and governance, mechanisms to halt or reverse interventions in the environment or individuals, a strong emphasis on equity and a desire to avoid monopolisation.

These findings will be of particular interest to research projects involving biotechnology, as they provide an initial, consolidated indication of the types of issues the public will want to see addressed when biotechnology of various kinds is deployed in real-world settings.

**Study of Arabidopsis thaliana COPT1 Transporter Function Using a Plant Membrane on a Chip Platform**

Authors:

Banuna, B. - Presenter

Huerta, M., Cornell University

Chia, J. C., Cornell University

Strakosas, X., Linköping University

Savvakis, M., Linköping University

Simon, D., Linköping University

Vatamaniuk, O., Cornell University

Daniel, S., Cornell University

A growing interest has been placed on the development of experimental techniques to understand the response of plants to climate change and stressors. The study of membrane proteins and transporters is integral towards understanding the key roles and functions of cellular processes in plant stressor response. Traditional techniques for investigating protein biophysics, such as the patch-clamp method, are less than perfect, lacking generalizability, ease-of-use, and speed. We propose a novel, fast, and sensitive biomimetic device for examining transport protein behavior based on a supported lipid bilayer (SLB). This platform comprises native plant membrane material on a transparent and conductive polymer (PEDOT:PSS), part of an organic electrochemical transistor (OECT). This allows the device to be amenable to a variety of optical and electrical characterization techniques. We have leveraged this platform to investigate the transport characteristics of a well-conserved copper ($Cu^+$) transporter protein COPT1 in *Arabidopsis thaliana,* responsible for maintaining copper homeostasis in plants. Mesophyll protoplast transiently expressing GFP and overexpressed COPT1:GFP were induced to form microvesicles. These microvesicles were reconstituted as a SLB using lipid-vesicles fusion on top of PEDOT:PSS. The formation and quality of the SLB was characterized by fluorescence recovery after photobleaching (FRAP) and confirmed by an increase in system impedance using Electrical Impedance Spectroscopy (EIS). Transport of $Cu^+$ through COPT-1 SLBs were measured by EIS under different concentrations 0-100 ÂµM. In the presence of $Cu^+$, COPT1 opens, allowing the transport of ions through the membrane, resulting in a decrease in the system impedance. Without $Cu^+$, the system impedance returns to the baseline SLB level pointing to system $Cu^+$ sensitivity. This indicates that our platform has the potential to be used to investigate plant transporters and proteins involved in response to plant stressors.

**Surface Electromyogram Neural Feature Based Gesture Classification Using Fastica**

Authors:

Kaleeswaran, R. - Presenter

Myoelectric control using pattern recognition methods is a critical research area in the field of rehabilitation engineering. Myoelectric control includes the data acquisition, signal processing, classification of movements and control of assistive devices such as exoskeletons and prosthetic arms based on the myoelectric signals. Gesture classification is a complex problem and needs precision. In this work, gestures are performed in various limb positions and EMG (Electromyogram) signals are recorded for multiple trials in each position. Independent Component Analysis (ICA) which is a Blind Source Separation (BSS) technique is used to decompose the motor unit action potentials from the recorded EMG signal. Feature space which consists of the extracted neural information, in addition to Root Mean Square (RMS) and Waveform Length (WL) of the EMG signal is utilized for classification of gestures using Support Vector Machine (SVM) classifier model. The results indicate that the proposed scheme is able to classify eight different hand gestures performed at five different limb positions with 98% accuracy.

**Synthetic Genetic Circuits to Modify Plant Root Structure**

Authors:

Brophy, J. - Presenter, Massachusetts Institute of Technology

Plant structure affects environmental stress tolerance. For example, the shape of a plantâ€™s root system influences its ability to reach essential nutrients in the soil or to acquire water during drought. Progress in engineering plant structure has been limited by our capacity to design and build genetic programs that alter growth in a predictable manner. We have built a library of synthetic transcription factors and used them to create synthetic genetic circuits that generate new spatial patterns of gene expression in Arabidopsis roots. Nicotiana benthamiana was used as a testbed to accelerate the design-build-test cycle of these circuits and enable productive tuning of circuit behavior. We apply the circuits to modify root structure by specifically changing root branching and gravitropism. This work highlights the utility of genetic circuits to enable precise spatial, temporal, and magnitudinal control over gene expression across entire organ systems and offers an exciting means to reprogram plant growth.

**Thermodynamics Plays an Important Role in Enhancing Terpene Productivity in Cyanobacteria**

Authors:

Shinde, S. - Presenter, Miami University (OH)

Singapuri, S., Miami University

Jiang, Z., Miami University

Long, B., Texas A&M University

Wilcox, D., Miami University

Klatt, C., Miami University

Yuan, J., Texas A&M University

Wang, X., Miami University

The terpene family is very diverse with more than 50,000 unique chemical structures. Terpenes are condensed from two $C_5$ precursor isomers, isopentenyl pyrophosphate (IPP) and dimethylallyl pyrophosphate (DMAPP). Limonene, a $C_{10}$ monoterpene, has been tested to use as a drop-in fuel additive for combustion engines and as the kerosene substitute for aviation fuels. Photosynthesis-driven $CO_2$ conversion to terpenes is particularly attractive for biofuel applications due to its low carbon footprint but is challenged with far lower productivities compared to heterotrophic hosts. We hypothesize that the low terpene yield in phototrophs is due to the low driving force determined by thermodynamics as it readily does not favor efficient photosynthetic $CO_2$ conversion. To bypass this limitation, we used the sigma factor engineering approach. An essential sigma factor, RpoD, encoding was overexpressed in photosynthetic cyanobacteria and showed a 1.7-fold improvement in limonene productivity. Further investigations on the engineered strains revealed a higher photosynthetic rate stimulating enhanced carbon fixation. Computational modeling and wet lab analyses showed an increased flux toward both native carbon sink glycogen synthesis and the non-native limonene synthesis from photosynthate output. Interestingly, our proteomics result shows the decreased abundance of terpene biosynthesis enzymes indicating their limited role in determining terpene flux. Thus, increased abundance of RpoD illustrates interesting metabolic rewiring that supports robust direct $CO_2$ to limonene conversion. This study reveals thermodynamics as the key determinant to boost low flux metabolites like terpene and demonstrates the merits of sigma factor engineering for efficient chemical productions.

**Uncovering the Biochemical Origin of the Causative Agents of Favism in Faba Bean**

Authors:

Geu-Flores, F. - Presenter

Faba bean (*Vicia faba* L.) is a widely adapted and high-yielding legume crop cultivated for its protein-rich seeds [1]. However, the seeds accumulate the pyrimidine glucosides vicine and convicine, which can cause haemolytic anaemia (favism) in 400 million genetically predisposed individuals [2]. Here, I will present our collaborative work aimed at uncovering the biochemical origin of these anti-nutrients, which was recently published in Nature Plants [3]. By using gene-to-metabolite correlations, gene mapping and genetic complementation, we identified a key enzyme in vicine and convicine biosynthesis, VC1. We showed that VC1 has GTP cyclohydrolase II activity and that the purine GTP is a precursor of both vicine and convicine. Finally, we demonstrated that cultivars with low vicine and convicine levels carry an inactivating insertion in the coding sequence of *VC1*. Our results reveal an unexpected, purine rather than pyrimidine, biosynthetic origin for vicine and convicine and pave the way for the development of faba bean cultivars that are completely free of these anti-nutrients.

[1] Duc, G. et al. in *Handbook of Plant Breeding: Grain Legumes* Vol. 10 (ed. De Ron, A.) Ch. 5 (Springer, 2015).

[2] Luzzatto, L. & Arese, P. Favism and glucose-6-phosphate dehydrogenase deficiency. *N. Engl. J. Med.* **378**, 1068â€"1069 (2018).

[3] BjÃ¶rnsdotter, E. et al. VC1 catalyses a key step in the biosynthesis of vicine in faba bean. *Nature Plants* **7**, 923-931 (2021).

**Advancing Biological Containment By Synthetic Auxotrophy in New Contexts and New Microbes**

Authors:

Kunjapur, A. - Presenter

Microbes play critical roles in mediating the health of soil and plants. Synthetic biologists have developed a variety of production, sensing, and response capabilities in living microbes; however, these tools currently cannot be implemented in the environment due to the risk of uncontrolled proliferation of engineered microbes. To address this issue, my lab has been working on designing strategies for intrinsic biological containment of microbes, with a focus on the technique of synthetic auxotrophy. A synthetic auxotroph is an organism engineered to depend on the provision of synthetic nutrients not found in nature. This poster highlights our latest characterization of Escherichia coli synthetic auxotrophs, which includes continuous evolution and direct addition to mammalian cell cultures. It also describes our successful efforts to engineer the incorporation of synthetic amino acids in the gram-positive microbe Bacillus subtilis. B. subtilis is a model system for both the study of bacterial cell biology and for industrial uses as a probiotic, rhizobacterium, efficient protein secretor, and spore-former. To enable greater understanding and control of proteins in B. subtilis, we achieved incorporated many distinct synthetic amino acids within proteins. We used these systems to achieve click-labelling, photo-crosslinking, and translational titration. Synthetic auxotrophy in E. coli and eventually in B. subtilis could serve as an enabling safeguard for engineered plant probiotics, particularly those with pre-programmed limited lifetimes in the environment.

**Characterization of Early Taxol Biosynthetic Pathway**

Authors:

Liu, J. - Presenter

Sattely, E., Stanford University

De La Peña, R., Stanford University

Taxol, a microtubule stabilizer derived from conifer *Taxus* species, is one of the most widely used anticancer drugs approved for treating multiple cancers. Current production methods include plant cell suspension culture and semi-synthesis, but both suffer from low yield, high production cost, and non-sustainability. Heterologous production of hard-to-source medicinal plant natural products in microorganisms has been shown to be successful in many cases. However, complete elucidation of the Taxol biosynthetic pathway would be a prerequisite for heterologous production. Despite decades of study on Taxol biosynthesis, the catalytic promiscuity of early biosynthetic enzymes, specifically the cytochrome P450 taxadiene 5Î±-hydroxylase (T5Î±H) that is attributed as the second enzyme on the pathway, has become a bottleneck in elucidating Taxol biosynthesis. When heterologously expressed in different organisms, T5Î±H has been reported to produce multiple oxidative products instead of a single product taxadien-5Î±-ol as expected. To understand if this catalytic promiscuity is inherent or due to extrinsic factors, we purified major products of T5Î±H when expressed in either *Nicotiana benthamiana* or *Saccharomyces cerevisiae* and elucidated their structures through extensive 2D NMR. We found that T5Î±H can perform up to three oxidations and at unique positions on taxadiene scaffold. The novel oxidation patterns observed in both *N. benthamiana* and *S. cerevisiae* suggests that this promiscuous catalytic behavior is likely inherent of T5Î±H. In addition, we found further conversion of some of these scaffolds into tailored products when co-expressed with potential Taxol biosynthetic genes. Some moieties of these structures resemble other taxanes discovered in *Taxus*, indicating a possible involvement of these enzymes to other taxanes. The diverse oxidative product profile paves the road for rational engineering of T5Î±H and suggests the presence of a complex metabolic network in Taxol biosynthesis.

**Characterization of Late Acting Enzymes in Neuroactive Lycopodium Alkaloid (HupA) Biosynthesis**

Authors:

Dho, Y. - Presenter

Nett, R.

Sattely, E., Stanford University

Plants have evolved to produce many diverse small molecules that affect mammalian nervous systems, making them a rich resource for treatments for neurological disorders. A prominent example is lysine-derived alkaloids, which are one of the least understood class of metabolites in plant biosynthesis. To better understand how plants synthesize these metabolites, we studied the biosynthesis of Lycopodium alkaloids that are produced by club mosses, focusing on Huperzine A (HupA)â€"a notable example of Lycopodium alkaloids known for its potential treatment for neurodegenerative diseases like Alzheimerâ€™s disease due to its acetylcholinesterase inhibitory effect. Through combination of metabolomics, transcriptomics, and enzymatic logics, we identified two novel 2-oxoglutarate dependent dioxygenases (2OGDs), non-heme Fe(II) bound oxidoreductase enzymes, in the late-biosynthetic pathway of HupA that do uncommon oxidative reactionsâ€"oxidative ring cleavage of HupB with a production of terminal olefin of HupC and irreversible double bond isomerization of HupC to HupA. To further study these remarkable novel chemistries performed by 2OGDs, the mechanistic details of these two 2OGDs were characterized *in vitro*. The first 2OGD alone was found to perform both oxidative ring cleavage and desaturation to form HupC from HupB through canonical catalytic cycle of 2OGD. The second 2OGD performed redox-neutral double bond isomerization on HupC to produce HupA using stoichiometric amount of 2-oxoglutarate (2OG). These mechanistic studies of ours on 2OGDs in the downstream of HupA biosynthesis expand the diversity of the known chemical transformations catalyzed by 2OGDs and offer deeper insights into how 2OGDs can perform complex chemistries, which could also potentially assist in identification and understanding of future enzymes involved in the biosynthesis of other lysine-derived bioactive alkaloids found in plants.

**Elucidating an Arabidopsis N-Responsive Gene Regulatory Network**

Authors:

Witham, S. - Presenter

Oz, T. - Presenter, DOE Center for Advanced Bioenergy and Bioproducts Innovation

Cai, Y. M., Earlham Institute

Demirer, G., UC Davis

Brady, S. M., University of California

Patron, N., Earlham Institute

Bian, C., East China University of Science and Technology

Nitrogen (N) is essential for plant growth and metabolic processes. The use of N-containing fertilisers helped to improve crop yield over the past few decades but had negative impacts on the environment. In response to changes in N availability, plants alter the expression of thousands of genes coordinated by a complex network of transcription factors. Previous work has identified a putative sub-network of transcription factors that are hypothesized to coordinate the large transcriptional response to changes in nitrate. However, the roles of the numerous TFs that comprise the network, many of which appear to work in feed-forward loops (FFLs) and to function epistatically, are unclear. In this project, we aim to characterise this sub-network, understanding how it coordinates plant growth in response to nitrate and to develop technologies that enable the rational engineering of N-use efficiency. Using *in vitro* and *in vivo* assays, we have investigated connections in the sub-network, identifying physical interactions and regulatory consequences (repression or activation). This has identified a key FFL that we hypothesise coordinates root growth in response to nitrate by modulating auxin signalling. Our results demonstrate how the components within this sub-network interact with each other and explain the phenotypes observed in loss-of-function mutations. It also provides a model for engineering plant N-responses using gene editing and network rewiring.

**Finding Molecular Targets for Engineering Auxin Signaling with Synthetic Interaction Domains**

Authors:

Bryant, J. Jr. - Presenter

Food security is in jeopardy due to climate change and the demands of population growth. A deep mechanistic understanding of auxin signaling proteins may allow rational crop engineering to accelerate improvements in agricultural efficiency. The auxin pathway is an ideal target for engineering since it regulates a plethora of developmental processes. Auxin is perceived by the TIR1/AFB receptor proteins which then bind to AUX/IAA repressors and relieve inhibition of ARF transcription factors to allow gene expression. We call this an auxin response circuit. However, the mechanism by which auxin signaling proteins regulate gene expression and physiology remains unclear. ARF transcription factor activity is typically regulated by interaction with AUX/IAA repressors through PB1 domains, which form high-order multimeric complexes. We hypothesize that the identity of ARF-Aux/IAA regulatory complexes determine promoter specificity and dynamics of gene regulation. However, this hypothesis is experimentally challenging to address due to the presence of numerous family members and the promiscuous, multimeric interactions between ARF and Aux/IAA proteins.

To untangle this complex regulatory pathway, we designed synthetic ARF and AUX/IAA constructs swapping the nonspecific PB1 interaction domains with highly specific interaction domains of tunable affinity using modular cloning methods. We expressed combinations of these constructs in yeast containing an auxin responsive promoter and TIR1/AFB auxin receptor to determine the impact of protein-protein binding specificity and affinity on gene expression dynamics. Steady-state activation assays performed on strains with only synthetic ARFs and an auxin responsive promoter showed elevated reporter activation which confirms that these synthetic transcription factors are functional *in vivo*. Strains containing the full auxin response circuit showed an auxin response similar to the native yeast ARC. Comparison of gene expression data between our ARF-Aux/IAA synthetic pairs vs. native pairs will allow us to refine our understanding of the mechanism by which auxin regulates numerous genes.

**Improvement of a Microbial Thiamin Biosynthetic Enzyme through Continuous Directed Evolution**

Authors:

Van Gelder, K. - Presenter

Garcia-Garcia, J. D., Escuela de Ingenier?a y Ciencias

Hanson, A., University of Florida

Thiamin is the essential cofactor for various key metabolic enzymes in all kingdoms of life. Thiamin is formed from a pyrimidine moiety and from a thiazole moiety derived from NAD, glycine, and a sulfur atom. In plants and yeast, thiazole is synthesized by THI4, a suicide enzyme that destroys an active-site cysteine residue by using it as the sulfur donor for the reaction and thus self-inactivates after just one reaction cycle. The resulting continuous replacement of THI4 is energetically costly. Certain prokaryotes have THI4s that use sulfide as a sulfur donor and are truly catalytic. However, these THI4s generally come from organisms from anaerobic, high-sulfide environments and their THI4s are consequently not expected to function well in plants. Therefore, we are using continuous directed evolution to evolve the THI4 from the anerobic bacterium *Mucinivorans hirudinis* to function in the relatively plant-like conditions (aerobic, low-sulfide) in yeast cells. Using the OrthoRep system in a yeast thiazole auxotroph reliant upon *M. hirudinis* THI4 for growth we applied three evolutionary strategies: 1) cold turkey - multiple passages without thiamin; 2) gradual â€" multiple passages in low thiamin followed by multiple passages without thiamin; and 3) variant library building â€" multiple passages in luxury thiamin followed by multiple passages with low thiamin, then without thiamin. These strategies led to selection of multiple single and double mutations in THI4 that often re-occurred in independent populations. These mutations were confirmed to confer faster growth by reintroducing them into a fresh vector in fresh yeast cells. The best of the mutant THI4s allowed growth rates close to that of native yeast THI4, leading us to deploy additional strategies to further increase the selection pressure; these strategies include downregulating expression of the target THI4 and reducing THI4 activity by adding an inhibitory thiazole analog to the selection medium.

# AUTHOR INDEX